Charles Gissler

# Contributions to the fauna of the New York Croton water

Microscopial observations during the years 1870-'71

Charles Gissler

**Contributions to the fauna of the New York Croton water**
*Microscopial observations during the years 1870-'71*

ISBN/EAN: 9783337273330

Printed in Europe, USA, Canada, Australia, Japan

Cover: Foto ©berggeist007 / pixelio.de

More available books at **www.hansebooks.com**

# CONTRIBUTIONS

TO THE

# FAUNA

OF THE

# YORK CROTON WATER.

## MICROSCOPICAL OBSERVATIONS

DURING THE YEARS 1870-'71.

By CHARLES F. GISSLER,

*...veloping state of Cypris...*

...TS AND FIVE PLATES, CONT... ...GRAVINGS ON STONE.

New York:

... STEAM PRINTER, No. 04 FULTON ST.

1872.

# INTRODUCTION.

In presenting this, my first endeavor, to you, I must beg your kind indulgence for a subject till recently but little studied ; if you, my reader, will only make an attempt, you will find it a most interesting study.

It is bound to but a small cahier, although the double volume would not be space enough to make a close and accurate description and explanation of only one single family of these little water-inhabitants.

It contains myriads of beings as perfectly organized as ourselves ; provided with organs, which, if not in as high a state of development as ours, nevertheless exist, and perform their functions instinctively, even if not guided by the distinguishing characteristic of man.

Many an hour have I passed with a tumbler of water in one hand and a loop in the other, always finding some new creature which till then had escaped my observation.

Let us remember L. AGASSIZ's words on the genus Coryne, and the same may be said of all the animalcules which occur in the New York Croton water. "In order to obtain a correct idea of this Hydroid, the observer must watch it in its native element, under all the circumstances and conditions of its natural mode of existence and development."

This pamphlet is accompanied by a few finely executed drawings, which, with but a few exceptions, are drawn

4

from nature by the author's own hand, and engraved on stone by Mr. F. Rixinger, of New York.

Also, seven wood-cuts, made by Mr. Bernstein, have been placed in the body of this pamphlet.

Of the orders and families having representatives in this water, there have been only one or two drawings made of each order.

Of the frequently found ALGÆ, fractures and single algæ, (*Conferva, Spirogyra, Desmidium, Diatoma, &c.*) I have had no drawings made.

The crustacean Cypris—figure is made after Carl Claus, Prof. of Zoology in Marburg ; (*Beitræge zur Kenntniss der Ostracoden*), some of the other figures are made superficially, representing only the outlines of their bodies. Many of these mentioned beings are only of periodical appearance, and for months may not be seen.

In closing, let me again urge upon you, my reader, how interesting you may make this study if you only wish. All that is required is a small MICROSCOPE, costing less than forty dollars, and a few other inexpensive implements.

Hoping that my labor may find a willing echo in your heart, I submit the following.

Respectfully,

February, 1872.                    *The Author.*

*"Certe natura nil facit frustra."*

The Croton water, after filtering by the very simple method adopted at the reservoir in the Central Park, contains, in addition to inorganic impurities, such as oxide of lead, carbonic acid, soda, magnesia, lime, and so forth, representatives of the vegetable kingdom, with their germs, (spores) and of animal life a very extensive group of Invertebrata and their larvæ in different stages of development. (See *.)

Of animalcules visible' only under the microscope, I found some of the class RADIOLARIA, Amoeba, Actinophrys, Acanthocystis and Difflugia. These belong to the lowest form of organized life, having no particular shape, consisting only of the so-called Protoplasma† or Sarcode, whose movements are few and very slow, remaining motionless for ten to fifteen minutes at a time, when it suddenly will elongate itself, contorting itself into all imaginable shapes, though naturally it assumes the globular form. Its food consists of putrescent corpuscles of Algæ and of single Algæ, (*Desmidians, Diatoms, etc.,*) which it devours by coming in contact with, and surrounding it, improvising a stomach in that part of its body which after digestion is ejected. The whole animalcule generally looks like a little transparent mass or jelly-like substance. Their movements and organization can only be distinctly observed under a high power of the

---

* Reports on the Water Supply of New York and Brooklyn.—Chemical Report by C. F. Chandler, Ph. D., Prof. of Anal. and Applied Chemistry, Columbia College, and Microscopical Report by Willliam B. Lewis, M D., New York. D. Appleton & Co., 1870.

† The Protoplasma, according to Huxley, is a compound of Oxygen, Water and Ammoniak, which view, Lionel A. Beal and T. H. Sterling contradict, and assert that it is impossible in any way to determine its composition.—*Fortnightly Review.*

microscope. (500—700 diameters.) The figure XII, on the third plate, represents the species DIFFLUGIA, found in the sediment of the Croton water. Another animalcule similar to an amœbial form, is the ACTINOPHRYS, remarkable for the long, tentacle-like, so-called " Pseudopodia," with which its globular body is surrounded, and which has not been fully represented on the wood-cut on account of want of space. I observed this interesting creature with another species Acanthocystis viridis, (*Grenacher in Z Z.*, 1869) during the July and August of 1870, and September of 1871.

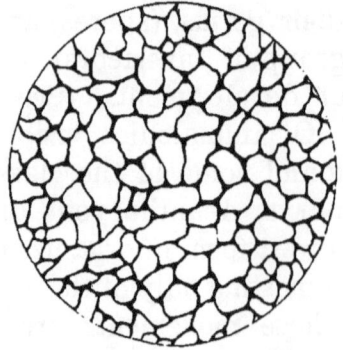

The class INFUSORIA, which are higher organized, are largely represented by the species Colpoda, Monas, Paramæcium, Cyclidium, Enchelys, Nassula, Chilodon, etc. Of more interesting Infusorial forms, I found VORTICELLA* CONVALLARIA, EPISTYLIS NUTANS, VOLVOX GLOBATOR, SPIROSTOMUM AMBIGUUM, whose movements and organization can be very easily observed with a magnifying power of 200 to 350 diameters. The VORTICELLA consists of an inverted bell-shape body, on whose open edges a number of vibrating fringes (cilia) in constant motion, are placed forming a sort of wreath around the top, which serve for locomotion and the drawing in of food, while at the lower surface is attached a leg-like projection consisting of a sheath with a central, longitudinal, muscular band, whose one end is fastened at the lower extremity, while the other end is fastened to the bell-shape body. This sheath

* W. Kuehne verified the *muscular* structure of the contractile band of the Vorticella-stem, by experiments made with the aid of electricity.

attaches it to any thing with which it may happen to come in contact, and the animalcule so attached swings in different directions, until a particle of food (*smaller Infusorials, single Algæ, vibrating Algæ—spores and so forth*) is brought into the mouth, and thence into the chymiferous cavity, when the stem contracts screw-like. Frequently, after water had been standing for hours, a perfect little tree of these Vorticells is formed, connected by their muscular stems.

Nine specimens have been drawn consisting of six different Infusorial species. CHILODON, SPIROSTOMUM, ENCHELYS, VORTICELLA, PARAMÆCIUM AND VOLVOX. The CHILODON drawing is accompanied with a figure, showing its successive division into two parts (the so-called semi-partition); the PARAMÆCIUM body is furnished with a star-shaped vesicle, called Vacuole, placed in the protoplasma of its body, altering its form every moment. This Vacuole is found in almost all Infusorials, and serves for locomotion; also for breathing as the ambulacral system of the Echinodermata or of the worms. The different forms (*A. Koelliker*) of the Vacuole or contractile space are shown on the third plate, at the figures XIV, a, ß, v, δ, ε.

Another organ in the Infusorial body is the so-called NUCLEUS, including the NUCLEOLUS. It is an integral part of the whole animalcule, taking part both in subdivision and in conjugation of any Infusorial in which it may happen to be. Almost all Infusorials are nucleated.

The increase of the Infusorials is very rapid, taking place by subdivision, and then scattering the body-cells in all directions throughout a large quantity of water.

Also, a kind of sexual fructification takes place in the Infusorials, whereby the Nucleus and Nucleolus play the part of the genital organs (Fertilization, Fecundation).

The type COELENTERATA is represented by a single

member, HYDRA VULGARIS (*plate IV and V, H*). The
HYDRA (*of 0,2 to 3,5 Mm. length*) is, besides the CORDY-
LOPOHORA LACUSTRIS (in England), the only polyp found
in fresh waters. After water has stood a few minutes in
a glass, this yellowish polyp seeks to fasten itself by its
lower extremity, which terminates in a little disk or
sucker, and then gradually expends its four to ten tentacles (feelers), waiting for food. If now any little animalcule comes in contact with one of its arms, they all con-
tract inwardly (*plate IV, H, VIII.*) and force the im-
prisoned animalcule into the mouth, which is placed in
the centre of the tentacles. There the nourishment is
brought into the alimentary canal (endoderm), and after
digestion is expelled through the mouth again. The tentacles are furnished with so-called nettle-cells (lasso-
cells), (*plate V at H, IX, and H X*), which instantly
kill the animalcules as they come in contact with them.
Owing probably to the presence of free formic acid,
which is supposed to be contained in the projected end
of the nettle-twine (in the lasso-cell). They increase*
by sexual fructification (male and female ovules reproductive calycle), by budding, and by separation (artifi-

---

* James D. Dana gives the following different modes of reproduction :

"I. Oviparous. 1. By ovules proceeding outward from the side of the
polyps, singly or in clusters. 2. By ovules formed from verticle lamellæ in
the visceral cavity and ejected through the mouth. The viviparous is but
an accident in the oviparous mode ; the eggs within develope in the same
manner as internally and for like reasons, as the external waters have free
admission.

"II. Gemmiparous. 1. By single buds, developing young, which after-
wards become free and independent animals. 2. By buds, which become
developed and remain persistent, and these may be either lateral or ter-
minal.

"III. By artificial sections. This mode may depend on the same cause as
the general distribution of the budding function, and may be properly an
analogous process, but depending on the imperfect character of the nervous
system, or its absence."

cial sections) of a single polyp into several distinct complete bodies. The sexual fructification takes place only in the months of September and October, as is the case with many Crustaceans (Daphnia, Sida, etc.) The Hydra vulgaris produces no hydro-medusarium, as almost all Hydroid-polyps do. (L. AGASSIZ).

Of the WORM type, the species NAÏS PROBOSCIDEA predominates, (*plate II, fig. VI, with natural length of* 1, 0 *to* 2,0 *Mm.*) which is very thin, with a still thinner proboscis. They reproduce very rapidly, by sexual fructification and by separation into parts, every fragment becoming a perfect animal, and the division may be continued indefinitely.

This phenomenon is caused partly by the multiplying of the body-cells and partly by the division of the body-segments (the so-called gemmipar and fissipar multiplication, (Carl Gegenbaur). The process begins always with the division of one metamere into several, which together form a complete, new individual.

In Naïs Proboscidea the advance in cellular differentiation of several segments was noted by me in specimens found in a little pond in the Central Park. The Naïs are found mostly in the sediment of the Croton water; their movements are wavy, snake-like.

The BRYOZOANS are found as greyish-white, branched pieces of a leathery consistence in the sediment of the water. The representative species being PLUMATELLA REPENS (*plate II, at the figure IX,* π), and CRISTADELLA MUCEDO. Viewed by a 70 to 100 diam. magnifying power, they offer to the eye a magnificent view by the movements of their tentacles (feelers), their manner of securing food, and other curious habits. They may be found in the summer-time in any of the small ponds of the Central Park on almost every stone, which they cover with their body-tubes (integumentum) like a moss. I saw the

Bryozoan CRISTADELLA MUCEDO only twice within the
two years I have been examining this drink-water. The
single ones are fastened together in three longitudinal-
rayed lines, with a general sucker on the lower part of
the body, which body moves but very slowly in the sediment. Of all Bryozoans the Cristadella mucedo is the
only one having the facility of moving. Allman, of
London, calls this Bryozoan-family the PHYLLACTO-
LÆMATA, on account of the tongue-like cover placed
over the ciliated mouth. The whole animalcule looks
like a very little transparent sponge.

Now and then we find a worm possessing character-
istics absolutely "sui generis" of a milky color, whose
stomach (*alimentary canal*) branches like a tree, whose
mouth is placed on the belly, having two eyes on the
head; this is called DENDROCOELUM LACTEUM, or PLAN-
ARIA LACTEA. (*Plate III, Fig. VII.*) Its body is very
soft, and surrounded with vibrating cilia, the movements
are snail-like. Its systematical position in Zoology is in
the order TURBELLARIA—Worms.

Three other creatures, belonging to the worms (ROTA-
TORIA)called ROTIFER, HYDATINA and STEPHANOPS, are
of frequent occurrence. They are very small organisms,
but larger than the largest Infusorials, possessed of a
mouth, jaws, intestine, chitinous skeleton, and increase
rapidly by the production of eggs.

Individuals with separate sexuality.

Of the Worm order, NEMATODA (*round worms*) occurs
ANGUILLULA (1,0 *to* 2,0 *Mm. long*), a worm occurring
in almost all fresh waters in the sediment, which have a
great resemblance to the very well known analogue, organized Trichina spiralis. Of the class CRUSTACEA are
found three orders, each represented by a few species.
The predominating CYCLOPS QUADRICORNIS belongs to
the order COPEPODA, has jumping (*saltating*) move-

ments, and generally swims by leap-like movements of the abdomen, but it wears its antennes in the position not as the fig V, a, plate II shows, but always in a horizontal position. The upward curved abdomen shows an obtuse angle with the cephalothorax. (*Plate II, fig. V, cth.*)

The CYCLOPS* has two pairs of antennes, (feelers) the abdomen has two separated, brush-like appendages, (*setæ*) one eye formed like an x, (the eye is drawn a little too wide on the figure) which is placed on the middle of the cephalothorax, the colorless blood is moved by the movements of the intestine ; the females are furnished during the whole year, except immediately after shedding, with an egg-bag, fastened on each side of the abdomen. (Genital-segment.)

The CYPRIS (0,7 *to* 1,8 *Mm. in length*), of the order OSTRACODA is greyish-brown, with two oval shields fastened together with a ligament on the back. This crustacean generally seeks the sediment of the water, and may be found always there. This creature has a single eye, (pigmented) which may be seen through its transparent, brown shields, placed on the forepart of its body.† The wood-cut represents a larva of the crustacean CYPRIS OVUM, in this state called NAUPLIUS. It is magnified three hundred times, and the drawing is made by a camera lucida with Hartnack's micr. from Paris, (*Claus.*) *A, the antenne of the first pair, five jointed ; B, the five jointed*

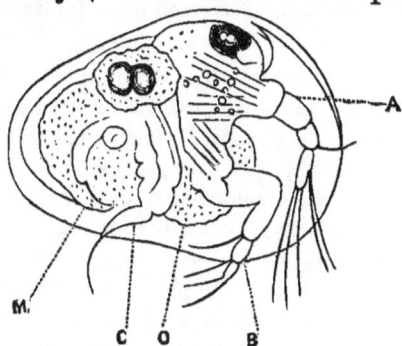

*antenne of the second pair; O, the maxillary part; C, the mandibular leg, four jointed; M, muscular band, fastened to the inside of the shields; the other organs are like those of the fully developed Cypris. (Fig. IV, plate* I.) The shields of Cypris become very transparent when touched with a drop of solution of caustic potassa. The male Cypris is very seldom observed, and the eggs are produced by the females as it is the case with Daphnia, Sida, the plant-lice, (Aphidæ), etc., by parthenogenetical production.*

The third crustacean occurring in this water, and also the largest, DAPHNIA PULEX (1,2 *to* 2,2 *Mm. long, and* 0,06 *to* 1,0 *wide*) of the order CLADOCERA or PHYLLOPODA, is reddish-grey with two scales. The much larger female carries her eggs under the scales on the back, until the embryos are in a forward state of development.† Seen under the microscope, they show a brain, a movable eye, surrounded with crystal-like, globular bodies, a mouth, simple intestine, anus, liver, heart, breathing apparatus, branched antennes, and so forth. The males only appear periodically in the autumnal months; during the inter-

---

plate X, 48 and 49. This species was found by Dr. Budd, in the neighborhood of Lake Champlain. Yet two other species are described by J. De Kay, C. agilis and C. simplex from the Lancaster County, Pennsylvania.

* Their relationship to the Poecilopoda, (Xiphosura) which are represented by the Limulus polyphemus (Horse-shoe) on the eastern coast of N. America, is very close. Van der Hoeven says in his work, page 37 (Recherches sur l'anatomie des Limules, Leyde 1838), "Mais soit qu'on range les Limules parmi les Crustacés, soit qu'on les mette avec les Arachnides, ils devront toujours former a eux seuls un ordre distinct, qui dans l'état actuel de nos connaissances, est éloigné de tous les autres ordres de ces deux classes. C'est en effet bien gratuitement et seulement d'après une simple ressemblance extérieure, que la plupart des naturalistes ont placé le genre Apus à côté des Limules. Leur (Phyllopoda) système nerveux diffère essentiellement de celu des Limules et consiste en deux séries de nombreux ganglions.

† The family Cyproidea is described in James D. Dana's Crustacean-work (U. S. Expl. Exped.,) pp. 1277 to 1304.

vening time, the females produce eggs by the above
mentioned parthenogenesis.

Of the crustacean Cyclops* and Daphnia,† I saw
still two other species, but so seldom indeed, that I could
not make any drawings of them. I consider them to be
CYCLOPS CORONATUS, (Copepoda) and SIDA CRISTALLINA
(Phyllopda). The Sida cr. has six rowing feet, (natatory
feet) its abdomen, with its appendages, being carried in a
horizontal position, one branch of the antennes is two,
the other three jointed. Some were two millim. in length,
and furnished on their backs with a sucker, with which
they fasten themselves to the glass, lying there somtimes
for hours, keeping their rapidly-moving rowing feet of
foliaceous character, in constant motion to assist breathing. (*These legs are like Daphnia's of branchial
character, gill-like. fr. Leydig.*) A few species of the
Cyclops coronatus (Copepoda) had the length of 1,03
Mm. Their anterior and posterior antennes have hairy
appendages. The Cyclops, the Daphnia,‡ and the
Cypris (no popular name has been given to any of these
little organisms) give amusement for hours by their constant jumps and leaps, as all four can easily be distin-
guished with the naked eye.

If several of these Crustaceans be kept in a little aqua-
rium with algæ, (*Conferva, Batrachospermum, etc.,*) they
may be observed for several months with their embryos,
eggs and manner of fructification. The Cyprids are
swarming in the swamps and ponds in the neighborhood
of the Central Park in millions, and they can be collected

---

* James D. Dana describes several Cyclops species in his work: C.
Brasiliensis, Curticaudatus, Pubescens, McLeayi, Vitiensis.

† In J. D. Dana's work, pp 1262 to 1275.

‡ De Kay places the Daphnians, after Cuvier, under the order Ostracoda,
and mentions a D. angulata and rotundata, found in stagnant water in the
forests of the Southern States.

by the handful without admixture with the four other
crustaceans (*Daphnia, Cyclops, Sida and Cypris*). The
PODOPHYRA CYCLOPUM (*Diesing*) is frequently found as
an ektoparasitical animalcule on the whole body of the
Cycl. coronatus, and C. quadricornis. It belongs to a
very simple organized family of the class Infusoria.
(ACINETÆ, SUCTORIÆ.)

By placing and pressing several specimens of Daphnia,
and also of the Cypris-larvae between two glass-covers
under the microscope, I found, but seldom during the last
Spring, (1871) a very interesting animalcule of a very low
organization, the parasite GREGARINA. The
wood-cut seems to show an Infusorial in divi- $\frac{400}{1}$
sion, but it has no contractile vesicle, (Vacuole) and no mouth, etc., only a nucleus-like
spot on each elongated side of its very elastic
body. The small wood-cut exemplifies two
Gregarines in Conjugation. Through this very conjugation and confoundation, the protoplasma of their bodies
is subdivided into innumerable little cells, called PSEUDO-
NAVICULÆ, by a kind of internal budding of cells from
the protoplasma. At a certain time, when these so
encysted Gregarines have grown larger, the external
cell-body (epidermis) breaks, and scatters the Pseudonaviculæ throughout the water, or as a guest in the
intestine of a Daphnia or Cypris, where the Gregarine
occurs. The minute organisms which serve as food for
other crustaceans, etc., grow larger in their intestines,
and the same process will begin again. Another wood-
cut has been made by way of comparison of a Gregarine, which generally is found in the tractus intetestinalis
of Blatta Americana (cock-roach) in the common earth
worm, also in almost all insects, especially in Coleoptera.
The encysted GREGARINA BLATTÆ has nearly the same
form as the encysted and conjugated Gregarine with

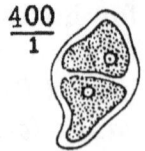

Pseudonaviculæ, which 1 found in the Daphnia. A very close examination into the development of the Gregarines has been made by Alexander Frantzius.* The systematical position is still somewhat doubtful; they generally are placed between the Radiolaria and Infusoria. By having opened hundreds of living cock-roaches, (not then having the fear of Mr. Bergh before me) I found thirty-three of these parasitical Gregarines in their intestines. Perhaps I may find in some future time, an occasion for continuing the examination of their development, etc., which will be published separately.

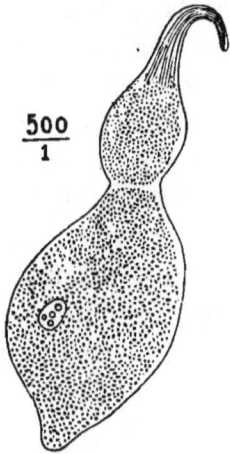

The easiest method for getting the larger animalcules swimming in this water (Crustaceans, worms, etc.) under the object-glass of a microscope, is the following : You fill a large-mouthed, half gallon bottle with fresh water, and let it stand for half an hour. If you have caught any one of these animalcules, you will see it swimming around, or having attached itself to the glass; take a glass-tube, having one of its ends drawn to a point, and draw the water just over the animalcule into the tube. Now, close the other end of the tube with the finger, and put the contents of the tube with the animalcule in a watch-glass, from which it is very easy to put under the microscope. The other smaller beings are generally only found by chance, or in the sediment, when the water has stood for some time.

These observations were all made with water collected in different parts of the city, and not from water taken from the reservoirs in the Central Park. Many of the mentioned beings are only of periodical appearance, and

* Observationes quaedam de Gregarinis. Vratislaviæ, 1846.

for months may not be seen. (*Hydra, Daphnia, Plumatella, etc.*)

I have never found any larvæ of insects, a not impossible circumstance, although so many millions are living in the reservoirs. I only wish to remind you of the myriads of Diptera, which on a fine Summer evening are swarming over the surface of the placid reservoirs, whose larvæ live in the water only, breathing by gills, after a certain time transforming into "pupæ," in which the "imago" is developed. The most interesting works on the development and histology of Diptera, are August Weismann's, from whose lips I imbibed the love of this interesting study.*

The little wood-cut represents the external part† of an abdominal segment of Cyclops with an encysted parasite, (600—700 diam.) which I several times observed. I regret that I had no time then to prosecute any further inquiries to determine its species, etc.

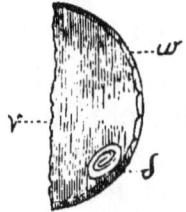

These little crustaceans, worms and polyps occurring in the Croton water for themselves alone, are not injurious, but when containing those encysted parasites may spread intestinal worms. (*Trematoda, Taeniada, Nematoda, etc.*

* Die Entwiklung der Dipteren, Leipzig, 1864, von Dr. A. Weismann, Prof. d. Zoologie, Metamorphose der Corethra pl., von A. Weismann, Prof. d. Zoologie in Freiburg in B.

† *Wood-cut Explanation.*—*w*, external chitinous body integument; *v,* fracture of the segment; *δ,* encysted parasitical trematoda larva (?)

N. B.—In the British Association for the Advancement of Science, held at London, September 14 to 21, 1870, was discussed the "Spontaneous Generation," and the "Spreading Agencies of Zymotic Diseases," by Huxley, Frankland, Child, Samuelson and Calvert. Samuelson finally expressed his opinion, resulting from experiments and observations which extended over a long series of years, that those who prefer to adopt the theory of the creation of living forms only from germs already in existence, would eventually find their view to be correct.—*I. London News.*

# LITERATURE.

DUJARDIN, histoire naturelle des Infusoires. Paris, 1841.

W. CARPENTER, researches on the Foraminifera. Philos. Transactions. 1856-'59.

C. G. EHRENBERG, die Infusorien als vollkommene Organismen. Leipzig, 1838.

FR. STEIN, die Infusorien auf ihre Entwiklung untersucht. Leipzig, 1854.

ALEXANDER ECKER, Hydra vulgaris. Freiburg i/B.

DUGES, recherches sur l'organisation et les mœurs des Planaires, annales des sciences naturelles. Serie I., Tab. XV.

OSCAR SCHMIDT, ueber dendrocœle Turbellarien. Zeitsch fuer wiss, Zool. X, XI.

ALLMAN, a monograph of the fresh water polyzoa. London, 1856.

BASTIAN, monograph on the Anguillulidæ. Trans. Linn. Soc., XXV. 1865. P. II.

American Naturalist, Salem, Mass.

IURINE, Histoire naturelle des Monocles. 1820.

FR. LEIDIG, naturgeschichte der Daphniden. Tuebingen, 1860.

CARL CLAUS, die freilebenden Copepoden. Leipzig, 1863.

W. LILJEBORG, Crustacea ex ordinibus tribus Cladocera, Ostracoda et Copepoda in Scand. occurentibus. Cum 27, tab aen. 1853. London.

F. DANA, Crustacea of the U. S. explor. exped., under Capt. Chs. Wilkes, 2 vol. and atlas, 1852.

JAMES DEKAY, Zoology of the State of New York, or the New York Fauna, Part VI, Crustaceans. Albany, 1844.

Besides the above mentioned books, consult the works of A. E. Verrill, Baer, Stimpson, Laurent, Cope, Smith, A. Hyatt, Shaw, Bailley. Stirling, etc., etc.

18

# EXPLANATION OF THE ENGRAVINGS.

PLATE I.—Fig. I, c. Chilodon cucullus, an Infusorial. Magnified about 650 diam. ; a, Nucleus with the Nucleolus : d, the mouth-opening, surrounded with a chitinous, fish-net-like apparatus, b, which is peculiar to the whole family of the Nassulina, c, oily, globular bodies; e, cilia, or hair-like fringes; ee, cuticula, cell-epidermis.

FIG. I, cc.—d, nucleus with nucleolus; b, mouth; a, fish-net-like apparatus; c, oily bodies; e, cilia. The Chilodon cucullus showing in division. Magnif. about 600—700 diam.

FIG. II.—Spirostomum ambiguum, an Infusorial. o, mouth; c, cilia; f, lower end, turned up; n. the chain-like nucleus with the nucleoli. Magnif. about 550—600 diam.

FIG. III. δ.—Daphnia pulex, a crustacean ( ♀ female). t, antennes; c, the brain; e, the eye, movable by two muscular bands; a, mouth; l, the liver; g, the so-called "scale-gland ;" p, the rowing-legs; q, two chitinous hair appendages (setæ); h, the heart; b, the "summer-egg-bag;" o, the eggs; i, the intestine; m, muscles; s, the two transparent scales; an, the anus on the abdominal end. Magnif. about 40 diam.

FIG. IV.—The crustacean Cypris fasciata in its sixth state of development. (The Cypris, according to Carl Claus, undergo nine changes in their development). oc, the pigment-eye; nc, nerve-centre; ov, the position of the ovarium; L, the liver; st, the position of the stomach; Ch, the first chitinous abdominal segment; π, protoplasmatical cells in the shields, which are connected with the hair-like appendages by very fine ramifications; g, the hind-leg under the "furca;" f, the three-jointed fore-leg: e, the maxillary leg; br, breathing apparatus (gill); m, muscles and ligaments; max., the maxilla; md., mandibula; mdf. mandibulary leg; 2 ant., the antenne of the second pair; sll. and sh., the two connected shields; bas. 2, p. ant., the basal joint of the second antenne-pair; 1, ant., the antenne of the first pair; S. D., the so-called shield-gland. The figure is made after C. Claus's work on the Ostracoda-developement, (Tab. II. fig. 17,) with closer explanation of the internal organs. Magnif. 135 diam. Seen from the right side.

PLATE II.—Fig. V shows the dorsal view of the Cyclops quadricornis, a crustacean, a, the first, b, the second pair of the antennes; c, the x like pigment-eye on the forepart of the cephalothorax; cth., the oesophagus:

$d$, seen from above; $f$, the liver-glands; $e$, the stomach, with $g$, the wavy moving intestine, around which the colorless blood circulates; $h$, the first of the four abdominal segments on whose underside the four pairs of rowing-legs can be seen; $p$, of which only the left-sided pairs have been drawn; $i$, the so-called "genital segment." The position of the genitals begins from the segment $h$, on both sides of the intestinal part $g$, extending to the segment $i$, consisting of very fine granulated, band-like glands, whose porus is on the female on both sides of the segment $i$, also carrying in that place the two egg bags; $K$. The much smaller and more seldom male-ones carry at certain times of the year the "Spermatophore," an elongated, bag-like body on the middle under-side of the segment $g$, which contains the products of the testicles, the spermatic particles to fructify the eggs of the females; $l$, the anus; $m$, the "furca," with the setaceous appendages $n$, magnif. about 60—65 diam. The wood-cuts represents a Spermatophore.

FIG. V.—$\varepsilon$, a young Cyclops with four legs (at a) around its mouth $b$., the oval body $d$, magnif. 300 diam. Fig. V.—$\varepsilon\varepsilon$, the same Cyclops in a higher state of development; $b$, the mouth, with the bag-like stomach; $c$, without anus; $a$, the eight legs with setaceous appendages drawn around the stomach, but not in favorable position for better showing the stomach—several segments are developed; $d$, the furca confounded with all other segments at $e$, and $f$, the furcal-brushes. Magn. 250 diam. The Cyclops-larva is generally called Nauplius, and undergoes five changes in their development. (Carpenter.) The wood-cut represents a Nauplius of a Cyclops.

FIG. VI.—The Naïs proboscidea (Annelide-worm); $p$, the tongue-like proboscis; $e$, the pigmented eyes; $o$, the œsophagus; $b$, a muscular bag, in which the proboscis $p$ can be retracted by the musculi retractores $m$; the bag is furnished on both sides with four brush-like bodies; $ch$., the chitinous brushes, (cirri) arranged in four longitudinal lines along the body; $t$, the chain-like stomach with dissepiments in constant motion for moving the colorless blood which surrounds it; $i$, the tractus intestinalis; and $a$, the anus. Magn. about 30 diam.

FIG. III. $\delta\delta$., shows the so-called ephippium, the egg-bag which the Daphnia carries during the winter time; $s$, the horny, chitinous epidermis of the bag, which is attached on the shields of the Daphnia by the ligaments m.; $c$, the glue-like mass in which the two hard-shelled winter eggs are placed.

FIG. IX. $\pi$., represents a piece of the Bryozoan Plumatella repensos, the mouth in the centre of the ciliated tentacles $f$, of which three are magnified 120 diameters at the fig. IX. $\delta$, with the cilia $c.$; $tb$, the inside ciliated funnel-body, retractile by the muscles $m$; $p$, the leather-like, elastic body-tube, with the buds at $b$; the fracture of the tube is shown at fr.; (a Plumatella, as the engraving IX. $\pi$. shows, I only saw three times, and I have not been fortunate enough to find it again); $\alpha$, the oesophagus; $s$, the stomach with the backward recurving intestine at $i$; the anus at an; the ovarium's position is at $ov$, with a gland-tube at $g$; the spermatic par ticles are produced opposite from the female apparatus at $ts$, with ciliated surface, and connected with the same gland-tube at $g$. Magnif. about 45—50 diam. The internal organs are only shown at one of the two branches (with closed tentacles). The two arrows show the direction of the movements of the tentacles. A Bryozoan-embryo, which moves very rapidly in the water, is represented by the fig. IX. $\beta$. Magnif. about 300—350 diam; $tb.$, a retractile tube with long cilia on the edges; $c$, the mouth with several very long hairs $fl$; $c$, cilia around the whole body; $pg$, pigment spots with longer cilia. Magnif. 360—400 diam. This figure has great resemblance with the figure of Dr. Lewis in the "Report on the Water Supply of New York and Brooklyn. Croton water sediment $i$. Halteria grandinella?

PLATE III.—Fig. VII.—A worm, Planaria lactea; $b$, the proboscis, expendable and retractile by muscles; $os$, the mouth; $t$, tree-like branched chimiferous channel (stomach); $e$, the two pigment-eyes, each with a globular, glass-like body (corpus vitreum?); $c$, the cilia surrounding the very soft body; $ov$, the two ovaria; $od$, the forked oviduct; $g$, the two testicles; $vd$, vasa deferentia; $ex$, genital-porus of this hermaphrodite.

FIG. VIII.—Hydatina senta, (a Rotatorial-worm). $os$, the mouth; $mx$, maxillary apparatus; $c$, the ciliated rotatory apparatus; $gl$, stomach-glands; $st.$, the alimentary cavity; $s$, segments around the chitinous body-integument; $ov$, the two ovaria; $e$, eggs; $ab$, ambulacral-tubes; $abc$, the two contractile vesicles of the ambulacral system; $an$, the anus; $p$, the furca. (†).

FIG. X., represents another animalcule belonging to the Rotatoria-worms (family Philodina), Rotifer vulgaris. $mx$, the chitinous jaws; $i$, the alimentary canals, with $gs$, two stomach-glands (secreting chimiferous juice); $oc$, two red-pigmented eyes on the forehead; edge, the two ciliated edges (two movable lobs); $ov$, the egg-producing cavity of protoplasmatical structure; $e$, two summer-eggs; $eb$, a nearly developed embryo; $ex$, the enlarged end of $i$; $an$, the anus; $ga$, intestinal glands (?), or rather a muscular band around $ex$, for removing the eggs when developed; $m$, muscles; $cd$, the spyglass-like appendage.

(†) After having become dry, they will revive on being moistened. The thin-shelled summer-eggs, whose position is on both sides of the alimentary canal, are produced by a parthenogentical manner; the thick-shelled wintereggs, fastened on both sides of the chitinous body near the furca (external), are fructified by the very small male-ones. (Enteroplea hydatinæ).

FIG. XI.—α and β, An Infusorial, Enchelys pupa, *os*, the ciliated mouth *s*, alimentary cavity; *an*, the anus. When young, colorless; older, greenish.

FIG. XII.—The Difflugia pyriformis (foraminifera, monothalamia). *s*, the protoplasma; *v*, the contractile vacuole with constant heart-like movements (Systole and Diastole); *sp*, the pseudopodia.

FIG. XIV.—α, β, υ, δ, ε, Exemplifying the 5, different states of the contractile space (vacuole). *sp*, space; *r*, rays: *a*, space contracted, rays filled; β, the first ejaculation of the rays, with filled space; *v*, the second time filled rays; δ, the second time emptied rays; ε, a moment before emptying the space, momentany showing the rays. (Koelliker).

PLATE IV.—Fig. XIII. β, an Infusorial, Vorticella microcostomum. *tr*, the screw-like contracted stem; *n*, Nucleus, including several nucleoli (Balbiani, Engelman); *v*, the vacuole; *os*, the mouth; *c*, the vibrating cilia (adoral ciliated peristome).

FIG. XIII.—ε, the same Infusorial in semi-partition; *tr*, muscular stem; *os*, the mouth; *n*, the semi-divided nucleus.

FIG. XIV.—The Paramaecium aurelia, an Infusorial, of which only the outlines have been drawn; *os*, the mouth; *v*, the contractile spaces; *c*, short, and *cc*, long cilia; *cp*, another contracted space, enlarged, showing the second state. (Plate III. Fig. XIV. β.)

FIG. XV.—The Volvox globator, an Infusorial. A globular jelly-mass, surrounded and having imbedded many individuals of Volvox, which are all connected together with a kind of proplasma-net. The larger balls in the figure XV, are, according to Cohn, the reproductive organs. Figure XV. *gg*, represents one of the small points by a high power of the microscope, showing two pseudopodia, several connecting links, and a few vacuoles in the protoplasma.

The Fig. H exemplifies the different forms and developments of Hydra vulgaris. Fig. H. VIII., a Hydra catching a Cyclops *cg*; *st*, the alimentary cavity; *tr*, the truncus body stem; *s*, the sucker or disk; fig. H. VII., a Hydra producing several persisting buds, *cd*; *f*, the tentacles. Magnifi. about ten diam. Fig. H. VI., a Hydre with two stem-buds *ld*; *os*, the mouth; fig. H. II., α, β, υ, three Hydres, each magnif. about 20—25 diam., in a more or less contracted state. Fig. H. V., an ovule (egg); *c*, epidermés. Fig. H. ββ., a horizontal tissue of a hydre-stem, with a bud *cd*; *pr*, the alimentary canal. Magn. 60—70 diam.

PLATE V.—Fig. H. III. *bd*, a young Hydre-bud, just appearing by expansion of the ectoderm; *eb*, a larger bud, a few days older, with four little tentacles, body ciliated; *tr*, the stem; *s*, the sucker; *os*, the mouth; *f*, the tentacles ; *st,* the alimentary cavity ; fig. H. IV.. *egg,* a porus egg under the ectoderm, fructified by the spermatic particles, which are contained in the little vesicles *sp ;* fig. H. IX., a 200 times magnified lasso-apparatus ; *c,* a vibrating cluster of smaller cells around a larger one ; *w,* the lasso-twine ; *n,* the oval cell, containing formic acid. Fig. H. X. represents the end of a tentacle with several lasso-twines ; *pr,* the streaming protoplasma with small granules.

♂ —The sign for a male.

♀ —The sign for a female.

☿ —The sign for a hermaphrodite.

---

### ERRATA.

At the wood-cut Cypris, page 11. the dotted points should reach from *m* to the little ring near the mandibular leg *c.*

# INDEX.

www.ingramcontent.com/pod-product-compliance
Lightning Source LLC
Chambersburg PA
CBHW022034190326
41519CB00010B/1711